War, Humanitarian Crises, Population Displacement, and Fertility

A Review of Evidence

Kenneth Hill

Roundtable on the Demography of Forced Migration
Committee on Population

NATIONAL RESEARCH COUNCIL
OF THE NATIONAL ACADEMIES

and
Program on Forced Migration and Health at the
Mailman School of Public Health
Columbia University

THE NATIONAL ACADEMIES PRESS
Washington, DC
www.nap.edu

THE NATIONAL ACADEMIES PRESS   500 Fifth Street, N.W.   Washington, DC 20001

NOTICE: The project that is the subject of this report was approved by the Governing Board of the National Research Council, whose members are drawn from the councils of the National Academy of Sciences, the National Academy of Engineering, and the Institute of Medicine. The members of the committee responsible for the report were chosen for their special competences and with regard for appropriate balance.

This study was supported by grants to the National Academy of Sciences and the Mailman School of Public Health of Columbia University by the Andrew W. Mellon Foundation. Any opinions, findings, conclusions, or recommendations expressed in this publication are those of the authors and do not necessarily reflect the view of the organizations or agencies that provided support for this project.

International Standard Book Number 0-309-09241-8 (Book)
International Standard Book Number 0-309-53284-1 (PDF)

Additional copies of this report are available from the National Academies Press, 500 Fifth Street, N.W., Lockbox 285, Washington, DC 20055; (800) 624-6242 or (202) 334-3313 (in the Washington metropolitan area); Internet, http://www.nap.edu

Copyright 2004 by the National Academy of Sciences. All rights reserved.

Printed in the United States of America

Suggested citation: National Research Council. (2004). *War, Humanitarian Crises, Population Displacement, and Fertility: A Review of Evidence.* Kenneth Hill. Roundtable on the Demography of Forced Migration. Committee on Population, Division of Behavioral and Social Sciences and Education and Program on Forced Migration and Health at the Mailman School of Public Health of Columbia University. Washington, DC: The National Academies Press.

THE NATIONAL ACADEMIES
Advisers to the Nation on Science, Engineering, and Medicine

The **National Academy of Sciences** is a private, nonprofit, self-perpetuating society of distinguished scholars engaged in scientific and engineering research, dedicated to the furtherance of science and technology and to their use for the general welfare. Upon the authority of the charter granted to it by the Congress in 1863, the Academy has a mandate that requires it to advise the federal government on scientific and technical matters. Dr. Bruce M. Alberts is president of the National Academy of Sciences.

The **National Academy of Engineering** was established in 1964, under the charter of the National Academy of Sciences, as a parallel organization of outstanding engineers. It is autonomous in its administration and in the selection of its members, sharing with the National Academy of Sciences the responsibility for advising the federal government. The National Academy of Engineering also sponsors engineering programs aimed at meeting national needs, encourages education and research, and recognizes the superior achievements of engineers. Dr. Wm. A. Wulf is president of the National Academy of Engineering.

The **Institute of Medicine** was established in 1970 by the National Academy of Sciences to secure the services of eminent members of appropriate professions in the examination of policy matters pertaining to the health of the public. The Institute acts under the responsibility given to the National Academy of Sciences by its congressional charter to be an adviser to the federal government and, upon its own initiative, to identify issues of medical care, research, and education. Dr. Harvey V. Fineberg is president of the Institute of Medicine.

The **National Research Council** was organized by the National Academy of Sciences in 1916 to associate the broad community of science and technology with the Academy's purposes of furthering knowledge and advising the federal government. Functioning in accordance with general policies determined by the Academy, the Council has become the principal operating agency of both the National Academy of Sciences and the National Academy of Engineering in providing services to the government, the public, and the scientific and engineering communities. The Council is administered jointly by both Academies and the Institute of Medicine. Dr. Bruce M. Alberts and Dr. Wm. A. Wulf are chair and vice chair, respectively, of the National Research Council.

www.national-academies.org

ROUNDTABLE ON THE DEMOGRAPHY OF FORCED MIGRATION
2004

CHARLES B. KEELY *(Chair)*, Walsh School of Foreign Service, Georgetown University
LINDA BARTLETT, Division of Reproductive Health, Centers for Disease Control and Prevention, Atlanta
RICHARD BLACK, Center for Development and Environment, University of Sussex
STEPHEN CASTLES, Refugee Studies Centre, University of Oxford
WILLIAM GARVELINK, Bureau of Humanitarian Response, U.S. Agency for International Development, Washington, DC
ANDRE GRIEKSPOOR, Emergency and Humanitarian Action Department, World Health Organization, Geneva
JOHN HAMMOCK, Feinstein International Famine Center, Tufts University
BELA HOVY, Population Data Unit, United Nations High Commissioner for Refugees, Geneva
JENNIFER LEANING, School of Public Health, Harvard University
NANCY LINDBORG, Mercy Corps, Washington, DC
CAROLYN MAKINSON, Andrew W. Mellon Foundation, New York
SUSAN F. MARTIN, Institute for the Study of International Migration, Georgetown University
W. COURTLAND ROBINSON, Center for Refugee and Disaster Studies, Johns Hopkins University
SHARON STANTON RUSSELL, Center for International Studies, Massachusetts Institute of Technology
WILLIAM SELTZER, Department of Sociology and Anthropology, Fordham University
PAUL SPIEGEL, United Nations High Commissioner for Refugees, Geneva
RONALD J. WALDMAN, Joseph L. Mailman School of Public Health, Columbia University
ANTHONY ZWI, School of Public Health and Community Medicine, University of New South Wales

Staff

BARNEY COHEN, *Director, Committee on Population*
ANA-MARIA IGNAT, *Senior Program Assistant*

COMMITTEE ON POPULATION
2004

KENNETH W. WACHTER (*Chair*), Department of Demography, University of California, Berkeley
ELLEN BRENNAN-GALVIN, School of Forestry and Environmental Studies, Yale University
JOHN N. HOBCRAFT, Population Investigation Committee, London School of Economics
CHARLES B. KEELY, Walsh School of Foreign Service, Georgetown University
DAVID I. KERTZER, Department of Anthropology, Brown University
CYNTHIA LLOYD, Population Council, New York
DOUGLAS S. MASSEY, Department of Sociology, Princeton University
RUBEN G. RUMBAUT, Center for Research on Immigration, Population and Public Policy, Department of Sociology, University of California, Irvine
JAMES W. VAUPEL, Max Planck Institute for Demographic Research, Rostock, Germany
ROBERT J. WILLIS, Institute for Social Research, University of Michigan, Ann Arbor

BARNEY COHEN, *Director*

Preface

In response to the need for more research on displaced persons, the Committee on Population developed the Roundtable on the Demography of Forced Migration in 1999. This activity, which is supported by the Andrew W. Mellon Foundation, provides a forum in which a diverse group of experts can discuss the state of knowledge about demographic structures and processes among people who are displaced by war and political violence, famine, natural disasters, or government projects or programs that destroy their homes and communities. The roundtable includes representatives from operational agencies, with long-standing field and administrative experience. It includes researchers and scientists with both applied and scholarly expertise in medicine, demography, and epidemiology. The group also includes representatives from government, international organizations, donors, universities, and nongovernmental organizations.

The roundtable is organized to be as inclusive as possible of relevant expertise and to provide occasions for substantive sharing to increase knowledge for all participants, with a view toward developing cumulative facts to inform policy and programs in complex humanitarian emergencies. To this aim, the roundtable has held annual workshops on a variety of topics, including mortality patterns in complex emergencies, demographic assessment techniques in emergency settings, and research ethics among conflict-affected and displaced populations.

Another role for the roundtable is to serve as a promoter of the best research in the field. The field is rich in practitioners but is lacking a coherent body of research. Therefore the roundtable and the Program on Forced Migration and Health at the Mailman School of Public Health of Columbia University have established a monograph series to promote research on various aspects of the demography of forced migration. These occasional monographs are individually authored documents presented to the roundtable and any recommendations or conclusions are solely attributable to the authors. It is hoped these monographs will result in the formulation of newer and more scientifically sound public health practices and policies and will identify areas in which new research is needed to guide the development of forced migration policy.

This monograph was prepared for and presented at the Workshop on Fertility and Reproductive Health in Complex Humanitarian Emergencies held in October 2002.

This monograph has been reviewed in draft form by individuals chosen for their diverse perspectives and technical expertise in accordance with procedures approved by the National Research Council's Report Review Committee. The purpose of this independent review is to provide candid and critical comments that will assist the institution in making the published monograph as accurate and as sound as possible. The review comments and draft manuscript remain confidential.

Ronald J. Waldman of Columbia University served as review coordinator for this report. We wish to thank the following individuals for their participation in the review of this report: Joan Kahn, Maryland Population Research Center, University of Maryland at College Park; Michael Toole, Macfarlane Burnet Institute, Melbourne, Australia; and Brad Woodruff, International Emergency and Refugee Health Branch, Centers for Disease Control and Prevention, Atlanta, Georgia.

Although the individuals listed above provided constructive comments and suggestions, it must be emphasized that responsibility for this monograph rests entirely with the authors.

This series of monographs is being made possible by a special collaboration between the Roundtable on the Demography of Forced Migration of the National Academies and the Program on Forced Migration and Health at the Mailman School of Public Health of Columbia University. We thank the Andrew W. Mellon Foundation for its continued support of the work of the roundtable and the program at Columbia. A special thanks is due Carolyn Makinson of the Mellon Foundation for her enthusiasm and significant expertise in the field of forced migration, which she has shared with the roundtable, and for her help in facilitating partnerships such as this.

Most of all, we are grateful to the author of this monograph. We hope that this publication contributes to both better policy and better practice in the field.

>   Charles B. Keely, *Chair*
>   Roundtable on the Demography of Forced Migration
>
>   Ronald J. Waldman, *Member*
>   Roundtable on the Demography of Forced Migration and Director, Program on Forced Migration at the Mailman School of Public Health of Columbia University

Contents

Introduction	1
Conceptual Framework	2
Likely Effects of Humanitarian Crises Through Intermediate Variables	4
Data Sources	5
Fertility Measures	8
Evidence of Immediate Fertility Effects	9
Evidence of Medium-Term Fertility Effects	14
Long-Term Consequences	22
Summary and Conclusions	24
Acknowledgment	26
References	26
Annex: Raw Numbers Underlying Figures 1 to 3	29
About the Author	32

War, Humanitarian Crises, Population Displacement, and Fertility: A Review of Evidence

INTRODUCTION

Fertility and reproductive health issues more broadly have tended to be of low priority in humanitarian crises (Wulf, 1994; Palmer et al., 1999). Public attention is drawn by information concerning the magnitude of refugee flows, of death tolls, and of numbers of injuries. Reproductive health has been regarded as a longer term issue that could safely be put on the back burner during the crisis phase of an emergency, when issues of providing adequate food, clean water, and shelter, plus treating acute infectious diseases of crowding, take priority. When reproductive health has been a priority, attention has focused on issues around sexually transmitted diseases and sexual violence. Although these are entirely appropriate priorities, little attention has been paid to the consequences for fertility, the major determinant of medium-term population dynamics, of humanitarian crises. The number of studies of fertility in refugee or displaced person populations has been very limited (Hynes et al., 2002).

In an important change for the international community, the International Conference on Population and Development (ICPD) in 1994, with its increased emphasis on human rights and reproductive health, explicitly included reference to reproductive health of both internally displaced persons and refugees and asylum seekers (United Nations, 1994). As a result of the increasing concern for the circumstances of refugees to which the ICPD recommendations were responding, more studies have become available in the past few years about fertility and reproductive health in displaced populations. However, concepts, data collection methods, and analytical methods vary widely, as do the circumstances of humanitarian crises (used broadly here to cover war, population displacements, famine, natural disasters, etc.), making comparisons and generalizations difficult.

The purpose of this paper is to review what evidence there is concerning the effects of humanitarian crisis on fertility, with a view to identifying common patterns that may exist across settings and be of value in guiding responses to future crises. I will start by adapting a conceptual framework for this purpose and reviewing data collection strategies and analyses.

CONCEPTUAL FRAMEWORK

The intermediate variables framework provides an appropriate structure for examining the effects of population displacement on fertility. As proposed by Davis and Blake (1956), reproduction is determined by three necessary processes: intercourse, conception, and parturition. As the authors state it: "In analyzing cultural influences on fertility, one may well start with the factors directly connected with these three processes. Such factors would be those through which, and only through which, cultural conditions can affect fertility" (authors' emphasis, p. 211).

Whereas Davis and Blake were interested in cultural factors, of interest here is the impact of humanitarian crises and population displacement, but the principle is the same: change in the intermediate variables is necessary and sufficient for fertility change, so any effect of crisis must work through one or more of these variables. Within each process, Davis and Blake identify the following variables (modified slightly for present purposes):

Process 1: Intercourse
- (a) Formation and dissolution of unions: age at entry into sexual unions, permanent celibacy, and amount of reproductive time spent between or after unions as a result of divorce, separation, desertion, or widowhood.
- (b) Exposure to intercourse within unions: voluntary abstinence, involuntary abstinence including temporary separations, and coital frequency.
- (c) Exposure to intercourse outside unions: coercive sex, commercial sex.

Process 2: Conception given intercourse
   (a) Involuntary infecundity (starvation, disease).
   (b) Use or nonuse of contraception.
   (c) Voluntary temporary infecundity (breast-feeding).
   (d) Voluntary permanent infecundity (sterilization, other medical procedures).

Process 3: Successful delivery given conception
   (a) Spontaneous intrauterine mortality.
   (b) Intentional intrauterine mortality.

Note that infanticide is not included. Demographers measure fertility in terms of live births, so infant deaths (and infanticide as a component of such deaths) do not directly affect fertility. It is possible that there may be some misreporting of infanticide as stillbirths in crisis situations, but evidence of such misreporting is nonexistent.

Bongaarts (1977, 1982) simplified the Davis-Blake framework into four intermediate variables associated with the vast majority of the variation of fertility between and within populations: (1) proportions currently married at each age weighted by marital fertility at that age; (2) proportions using contraception of varying levels of effectiveness; (3) postpartum infecundity associated with breast-feeding or postpartum abstinence; and (4) the extent of induced abortion. Bongaarts (1982) demonstrated that the remaining factors either do not have a large effect on fertility (e.g., spontaneous intrauterine mortality) or do not vary much between populations (e.g., permanent sterility). However, Bongaarts' arguments are for fertility in a "business-as-usual" setting, whereas humanitarian crises are anything but business as usual. I will thus retain the broader range of intermediate variables proposed by Davis and Blake, in order to be sure that abnormal circumstances are covered.

The key characteristics of intermediate variables are that they are both necessary and sufficient for fertility change, and that the direction of change in fertility resulting from a change in an intermediate variable is unambiguous. Thus any background, cultural, or other factor must operate through one or more of the variables, and it must have an effect on fertility whose direction (if not size) through that variable is unambiguous.

LIKELY EFFECTS OF HUMANITARIAN CRISES THROUGH INTERMEDIATE VARIABLES

The effects of humanitarian crises on fertility are likely to depend to some extent on the stage of demographic transition reached by the population. Humanitarian crises vary widely in nature and setting, but they will be characterized by the affected population experiencing one or more of the following adverse consequences: direct exposure to violence, witnessing violence, loss of family members, displacement, food scarcity, increased exposure to communicable diseases, reduced access to health services, plus a range of other socioeconomic setbacks. Effects on fertility will depend on the adverse consequences experienced and on the characteristics of the affected population. In a largely natural fertility population (that is, a population whose fertility-related behaviors do not vary substantially by achieved family size), effects on fertility can be expected to be largely involuntary, except perhaps for changes in the incidence of induced abortion. In a population with a substantial level of fertility control, in contrast, changes in use of contraception in response to changes in fertility preferences (concerning both timing and ultimate family size) resulting from the crisis may be substantial.

In the short run, effects through intercourse are likely to reduce fertility. A humanitarian crisis is likely to delay entry into a sexual union and to increase the risk of spousal separation or union dissolution, with a downward effect on fertility. A crisis is unlikely to have much effect on permanent celibacy. Within unions, involuntary abstinence through temporary separation is likely to increase, and in severe cases coital frequency may be reduced, again exerting a downward effect on fertility. Intercourse outside unions may increase, however, either through coercion or commercial sex, with the possibility of increased extramarital childbearing.

Effects through risk of conception are less clear-cut. Although moderate malnutrition appears to have little effect on fecundability (Bongaarts, 1980), starvation clearly reduces conception probabilities, sharply reducing fertility in serious famines (Stein et al., 1975). Fecundability may also be affected by sexually transmitted diseases (including HIV) and certain other infections, such as malaria, the prevalence of which may increase in crowded camp conditions. In populations with substantial voluntary fertility control, use of contraception may be interrupted by a crisis, but couples may also choose to avoid pregnancy during difficult times. Breast-feeding may be continued for longer periods, increasing postpartum amenorrhea, but it may be truncated by increased risk of infant death. Although women of reproductive age appear to have an above-average propensity to move in response to a crisis, young children are sometimes underrepresented in camp populations (Holck and Cates, 1982), either because they never started a move or because they died en route.

Although data concerning risks of intrauterine mortality are very weak, there is no compelling evidence that malnutrition or stress has a substantial effect on risks of spontaneous pregnancy loss. Induced abortion, in contrast, might reasonably be expected to increase during a crisis if abortion is within the nexus of conscious choice, thus reducing fertility.

In the long run, the directions of effect become even less clear. Long-term refugee populations may well marry early, because of lack of attractive alternatives, and may prefer large families for political reasons; for example, fertility has remained very high in the Palestinian population over several decades (Fargues, 2000). Couples may also react to long-term insecurity either by increasing childbearing (to make up for child losses, as a source of future security, or to maintain political influence) or by reducing it as a result of pessimism about the future.

DATA SOURCES

Statistically developed countries obtain information about fertility levels and trends from virtually complete birth registration. However, registration of births in many developing countries is not complete, and a humanitarian crisis is likely to have an additional adverse effect on an existing registration system. Thus civil registration data are not generally useful sources of data about the fertility effects of severe crises (although they may be useful for studying the effects of economic fluctuations on fertility).

Demographic data for refugee or displaced populations used to come largely from camp registers. However, over the past two decades, the use of data from household surveys has become much more common. Such surveys have been conducted in camp populations, in self-settled refugee populations, and in nationally representative samples. Regardless of source, however, such errors as misreporting of age, of dates, or even wholesale omission of events often affect data quality in developing countries.

The accuracy of data from camp registration systems is likely to vary for a variety of reasons, among them number of staff, duration of existence, and the nature of entitlement systems. A recent review of reproductive health indicators in 52 postemergency phase camps in seven countries found that the crude birth rate was positively associated with the number of local health staff per 1,000 persons and negatively associated with the age of the camp (Hynes et al., 2002). Data quality may also be affected by the services offered. If food and other entitlements are linked to registers, there are risks of multiple registrations of residents to receive increased resources, of inclusion of persons who are not true residents, and of omission of events that remove persons, such as deaths. Births are likely to be fully or more than fully recorded, but the size of the resident population may be exaggerated. If entitlements are not linked to the register, events (such as births at home) that occur outside a facility are likely to be underrecorded, as is the resident population. Bias is possible in both cases, although the direction of the bias is not clear. An additional problem with register data is limited content: the age distribution of the population has in some cases been recorded or presented only in the broad categories of under age 5 and age 5 and over, a situation that does not permit adequate age standardization of crude rates.

Retrospective surveys have collected data on fertility primarily by asking women about their own childbearing experience. The simplest question simply asks each woman about the number of children she has given birth to. It is well established that omission of births, perhaps largely children who died before the interview, is a common problem, particularly among older women (Brass et al., 1968). Also, such an aggregate birth history provides no information about the timing of births and is thus of no value for tracking short-term changes in fertility. Current fertility has often been estimated from a question about recent childbearing—either whether or not a woman had a birth in the 12 months before interview or for the date of her most recent live birth. This approach has uniformly been found to underestimate fertility, either because of a misperception of the time period involved or as a result of reporting events as occurring further in the past than was actually the case (United Nations, 1983). Also, data from a single survey provide no information about changes over time or in response to a crisis.

In order to get more information, some surveys have used a "truncated" birth history, collecting information about all births in a defined time period (for instance, the 5 years before the interview) or about some number of recent births, such as the last three births. The time-truncated birth history has been found to underestimate fertility greatly, and birth histories truncated by some number of births are likely to be affected by errors of date recording that are also likely to reduce recent fertility. Full-birth histories, whereby a woman is asked for the date of birth of each live-born child (and age at death if he or she has died), is the method of choice for fertility surveys today and is used by all Demographic and Health Surveys (DHS). Even such full-birth histories, however, are affected by errors of date reporting, which may shift births from one time period into another (Brass and Rashad, 1992). The DHS have noted a tendency to shift births out of the time period covered by extra questions about the health of young children (Arnold and Blanc, 1990), and fertility trends from birth histories have been found often to exaggerate fertility declines, as a result of reporting births further in the past than they actually occurred. That said, full-birth histories are not much more time-intensive in terms of field work than, say, a truncated history focusing on the last three births, but they allow much more flexibility in analysis and provide a much stronger basis for internal consistency checks to identify possible errors.

An additional problem with birth histories, whether aggregate, truncated, or full, is that they are obtained from interviews with surviving women. A woman who has died or moved out of a sampled area cannot provide information. Thus the data collected will reflect any sample selection biases: survivor bias may be expected to minimize the severity of the effects of a crisis on fertility. Although there is no straightforward way around this problem, an indirect approach, based on asking women about the survival or otherwise of their sisters, and an aggregate pregnancy history for each sister, might provide an indication of the magnitude of any selection bias. However, this approach has not been tried in the field.

In some cases, the only information available about a population is its age distribution. An approximate measure of fertility can be calculated as the child-woman ratio, the ratio of children under 5 to women ages 15 to 44. This measure may be affected by high child mortality and also by typical patterns of age misreporting (United Nations, 1983).

The bottom line is that there is no universally satisfactory way of collecting information about fertility during a humanitarian crisis. This is one reason why relatively little is known about the effects of crisis on fertility, and a further reason for treating empirical results with some caution. Retrospective measures, in particular, are subject to biases in terms of event dating that may create spurious fluctuations in numbers of births over time; much greater confidence in time patterns can be felt if two different surveys, held at different times, show similar fluctuations for the same time periods.

FERTILITY MEASURES

Demographers are obsessive about two things: age and exposure time. We obsess about age because demographic processes are highly age-dependent, with the result that quite small differences in age distributions can have a substantial effect on summary measures, such as the crude birth rate. Failure to take account of age distributional factors can give rise to highly misleading comparisons. We obsess about exposure time in part because we obsess about age (and thus need to get exposure to risk in different age ranges correct) and partly because demographic events take place over time, so that roughly speaking twice as many events take place over a 1-year period as over a 6-month period. Controlling for exposure time is thus a critical part of any meaningful comparison.

As a result of these two obsessions, demographers tend to describe processes in terms of age-specific occurrence-exposure (O-E) rates, whereby numbers of occurrences (pregnancy outcomes and particularly live births, for this paper) to women of a particular age or age group during a time period are related to the total exposure time of women at that age or age group during that time period. The measure of exposure time is usually expressed in person-years, and using this standard is convenient for the purposes of comparison with, for example, more normal conditions. However, the choice is arbitrary. It is important to note that the measure of exposure time used has no necessary connection to the length of time over which measurement lasts: observation over 10 days results in 0.0274 person-years of exposure, and results can perfectly well be presented on an annualized basis.

In situations in which age-specific O-E fertility rates can be calculated, the summary fertility measure of choice is the Total Fertility Rate (TFR). The TFR represents the average number of children a woman would have if exposed throughout her reproductive life to a given set of age-specific fertility rates. It is calculated by summing such fertility rates across all ages of childbearing, and is unaffected by the age distribution of the population.

It is often the case that the data available simply do not permit the calculation of age-specific O-E rates. Births may not be recorded by age, or they may be recorded in terms of very wide age groups. If no age information is available for either births or population, the crude birth rate (CBR), calculated as births in a period divided by the midperiod population multiplied by the length of the period in years (in most cases an adequate approximation for exposure time), is all that can be calculated. It must be remembered, however, that the birth rate may be affected by the age and sex distribution of the population. If some information is available on the age and sex distribution of the population (but not for births), a somewhat more specific measure, the general fertility rate (GFR), can be calculated by dividing the births by the female population ages 15 to 44. In normal populations, the GFR is quite insensitive to typical variations in age distributions, but displaced populations are likely to be abnormal, and the GFR needs to be examined carefully. For comparisons between populations, indirect standardization using the full age distribution is preferable to the GFR, and if any age of mother detail at all is available for births, direct standardization (as described, for example, in Preston et al., 2001) is better still.

A current estimate of GFR that may be less biased by date of event errors than estimates based on retrospective reports of recent births can be derived from information on current pregnancy. All (ever-married) women of reproductive age can be asked whether they are currently pregnant. The proportion reporting a current pregnancy is typically too low, possibly because of unrecognized pregnancies, but the degree of underreporting seems to be fairly constant across settings (Stanton, 2004). Figure 1 plots proportions currently pregnant (PCP) against the GFR for the preceding 3 years for 114 DHS around the world. The association is clearly quite close, and the regression line (which accounts for 99 percent of variance)

$$GFR = 24.30*PCP - 0.492*PCP^2$$

provides a simple way to obtain a current fertility measure. It is interesting to note that the relationship is slightly curvilinear: reporting of current pregnancy may be slightly better, or pregnancy loss somewhat lower, in low-fertility than in high-fertility populations.

EVIDENCE OF IMMEDIATE FERTILITY EFFECTS

A crisis or displacement can have a short-term effect on fertility only through the third (gestation) process of the conceptual framework, since pregnancies will already have occurred. Process 3 has only two variables, spontaneous and intentional intrauterine mortality. The majority of the evidence concerning spontaneous pregnancy loss is from famines and suggests that there is little effect. Mosley (1978) cites evidence from both The Netherlands in 1944-1945 and Bangladesh 1974-1975 that severe famine had no measurable effect on spontaneous pregnancy loss, although the data

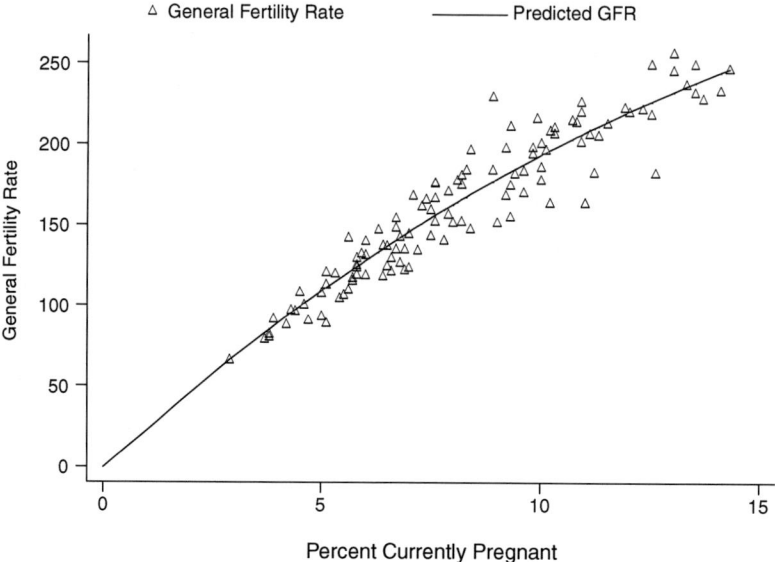

FIGURE 1 Relationship between general fertility rate (in 3 years before survey) with proportion currently pregnant (at time of survey) for 114 DHS.
SOURCE: Stanton (2004).

are not strong. Intentional intrauterine mortality could have a very large effect, but there is very little evidence on this issue.

Mosley (1978) cites Chowdhury and Chen (1977) as showing that there was a significant increase in induced abortion in Matlab Thana during the Bangladesh famine of 1974-1975, in a population in which induced abortion at the time was probably quite unusual. The monthly data on pregnancy outcomes in Matlab, collected by regular household visits by field workers, represent perhaps the only prospective data source concerning monthly miscarriages, stillbirths, and live births during a famine in the developing world. The pattern of miscarriages in comparison to births does not seem to provide strong evidence for changes in overall pregnancy loss.

Figure 2 shows the monthly deviations from the seasonally adjusted trends in live births, stillbirths, and miscarriages over the period January 1974 to December 1977 for each outcome. The price of rice rose throughout 1974, peaking in the first quarter of 1975 at over 200 percent of initial levels (Mosley, 1978) . As can be seen, live births and stillbirths show very

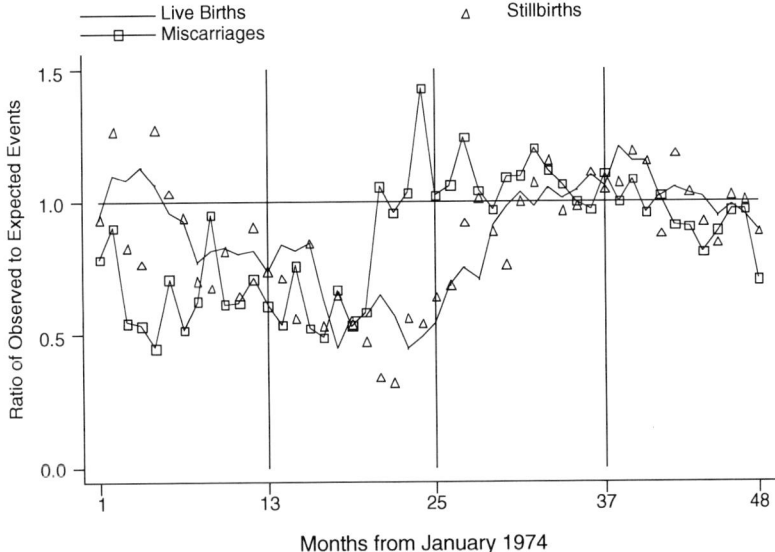

FIGURE 2 Monthly deviations in live births, stillbirths, and miscarriages from seasonally adjusted trends, Matlab, January 1974 to December 1977.
SOURCE: Cholera Research Laboratory. Demographic Surveillance System—Matlab. Registration of Demographic Events, Reports 4 to 7, 1974 to 1977, Dhaka, Bangladesh.

similar patterns, both dipping below expected values from August 1974 and remaining low until May 1976. If the decline in births was the result of reduced conceptions, the apparent drop by August 1974 is surprisingly early, but the more pronounced drop in May 1975, continuing to January 1976, is consistent with reduced conceptions at the peak of the famine. Miscarriages, presumably including induced abortions in these data, fall well below expected values as early as March 1974 and remain low until September 1975.

These data certainly give no support to the idea that miscarriages might have increased as a result of the famine, and it is not until September 1975 that they give any support to the idea that induced abortions might have increased (but the fact that miscarriages increased to "normal" levels by September 1975, 9 months before births recovered, may suggest some induced abortion in the later stages of the famine). The pattern of miscarriages both falling and then recovering before births suggests that it was changes in conceptions that were responsible for both.

Evidence from other developing country settings is lacking. However, an undated report of the Technical Working Group on Reproductive Health and Pregnancy Outcome Among Displaced Women of the International Centre for Migration and Health describes the reproductive health situation of displaced and local women in Sarajevo before and during the Bosnia war in 1992-1994. The absolute number of deliveries in Sarajevo fell from about 9,000 in 1988 and 1991 to only slightly above 3,000 in 1992, and as low as 2,000 in 1993 and 1994. Part of the effect is ascribed to the evacuation of a number of pregnant women during the initial months of the war, as well as to the fact that many couples were split up and remained separated for long periods of time (Process 1(a)). However, the ratio of induced abortions to live births doubled during the war, from one abortion per pregnancy taken to term prior to the war to at least two during the war. Contraceptive use even before the war was very low—only 5 percent of women reported ever having used a contraceptive device or pill—so the main form of birth control had been abortion. The report also notes a steep rise in the proportion of low-birthweight babies during the war, from 5 percent before the war to 13 percent during the war: it is possible that the increase in undernutrition reflected in this increase may have been large enough to reduce fecundability to some extent. Thus fertility in Sarajevo fell as a result of Process 1(b) (temporary separation) and Process 3(b) (abortion) and may also have been affected by Process 2(a) (reduced fecundability).

One of the few studies that look at displacement rather than famine presents data from two Cambodian refugee camps in Thailand following the Vietnamese invasion of Cambodia and the collapse of the Khmer Rouge regime (Holck and Cates, 1982). Two holding camps were established, one (Sakaeo) that received refugees from mostly rural and low socioeconomic settings, and the other (Khao I Dang) that received refugees from mainly urban backgrounds and of higher socioeconomic status. Data on births, deaths, and hospital admissions were obtained through a registry, and periodic surveys were conducted to estimate the age and sex distribution of the population. The two camps had markedly different fertility levels in the 6-month period from November 1979 to April 1980: the CBRs were 13 and 55 per 1,000 in Sakaeo and Khao I Dang, respectively, and GFRs were 35 and 203 per 1,000 women ages 15 to 44. Fertility in Khao I Dang was similar to the fertility level in Cambodia prior to the crisis, but fertility in Sakaeo was as little as one-quarter of its precrisis level. Percentages of women 15 to 44 who were pregnant support the differences: 3.2 percent in Sakaeo,

11.3 percent in Khao I Dang.[1] Given the intermediate variable framework, it is highly unlikely that these differences reflect short-term effects of displacement. It is much more likely that they reflect lower conception rates prior to arriving in the camp in the rural-origin population of Sakaeo than in the urban-origin population of Khao I Dang. The rural-origin women delivered babies with substantially lower birthweights (27 percent below 2,500 grams) than the urban-origin women (20 percent below 2,500 grams), and over 80 percent of the women arriving at Sakaeo were amenorrheic or dysmenorrheic. It seems overwhelmingly likely that the women in the Sakaeo camp were severely malnourished before arriving in the camp.

This study raises a number of interesting methodological issues. The authors note that their registry included only births that came to the attention of the hospital staff, and they conclude from survey data that since as many as 25 percent of births did not occur in hospital, underrecording could be substantial (Holck and Cates, 1982). The authors also note that the age distributions of the two populations were rather different from that of the general Cambodian population, with lower proportions of both children (about 6 percent in Khao I Dang and 4.5 percent in Sakaeo under age 5, compared with over 9 percent in the 1962 population census of Cambodia) and adults over age 45. The age distributions of women of reproductive age in the two camps were also quite different, making comparisons of nonstandardized measures such as the CBR and the GFR problematic. The paper states that more than 60 percent of the women in Sakaeo were under age 25, but fertility measures were not age-standardized because only two age intervals, 15-24 and 25-44, were available. Given the very different age distributions, even a two-age group standardization would have been helpful.

Other interesting findings from this study include the fact that 15 percent of married women were not with their husbands in Sakaeo, a factor likely to depress subsequent fertility through the exposure to intercourse process of the conceptual framework. Only 47 percent of the women ages 15 to 44 in Sakaeo had been married, probably largely reflect-

---

[1] It is interesting to note that these percentages pregnant predict a GFR value, given the model above, that is closely consistent with the observed value for Khao I Dang (212) but is double the observed value for Sakaeo (77). It may be that improved nutrition after arrival in the camp had a rapid effect on increased pregnancy rates in Sakaeo.

ing the age distribution, but another factor likely to explain differences in unstandardized fertility measures between the two camps.

A subsequent paper (Centers for Disease Control, 1983) reported on the population of Khao I Dang camp for the period December 1981 to November 1982. During this later period, the CBR remained close to 55 per 1,000, while the proportion of low-birthweight children fell to 9 percent. That fertility changed little, even as the proportion of low-birthweight children fell by half, supports the idea that the short-term effect of the displacement on fertility was low.

EVIDENCE OF MEDIUM-TERM FERTILITY EFFECTS

Medium-term effects can work through all three processes of the conceptual framework, although some components are unlikely to play a significant role. Since spontaneous pregnancy loss appears to be relatively unimportant immediately, it is unlikely to play a medium-term role. Permanent celibacy is also unlikely to play a medium-term role.

Historical studies clearly show that fertility covaries with economic conditions (Galloway, 1988; Lee, 1990), and some contemporary studies show similar effects (Palloni et al., 1996). Here, however, we are interested in the fertility effects of more catastrophic changes: famines, civil wars, and mass population displacements.

There is a substantial literature on the fertility effects of famines. Stein et al. (1975), using civil registration data plus hospital and military records, report a drop of up to two-thirds in fertility 9 months after the Dutch famine of October 1944 to May 1945. Comparisons with control areas support the conclusion that it was the famine, rather than climate or psychological stress that was responsible. Detailed analysis of the impact by age, parity, and social class led the authors to conclude that the effect was probably on fecundability (the probability of a conception per month given unprotected intercourse) rather than through reduced sexual activity or increased fetal loss. Birthweight was also affected by the famine, particularly for pregnancies in their third trimester during the famine. Fertility recovered immediately after the famine, although the rebound to levels higher than those that preceded the famine was also observed in the control areas.

Various reconstructions of the famine in China in the late 1950s all show major reductions in fertility; for example, Ashton et al. (1984), using a combination of census and civil registration data, estimate that the CBR fell in 1960-1961 to only 50 percent of its value in the early 1950s (rebounding about 10 percent above normal in 1962-1963). The CBR in Matlab Thana, Bangladesh, as recorded by an intensive surveillance system, declined by about 10 percent in the year after the Bangladesh civil war (Curlin et al., 1976) and by about 40 percent from its prefamine level between April 1975 and March 1976 (Chowdhury and Curlin, 1977, quoted in Mosley, 1978), although Figure 2 suggests a 20 percent reduction as early as August 1974 and a peak reduction of 50 percent from May 1975 to January 1976.

Only limited information is available about the medium-term fertility effects of humanitarian crises. A variety of data sources and methodologies have been used. Khlat et al. (1997) used birth registers in hospitals to estimate fertility levels in Beirut during the Lebanese civil war and subsequent economic hardships. Specifically, they use data on the age of mother and birth order of deliveries in 1984 and 1991 by religion and class of hospital with an indirect analysis method developed by Brass (1969) and Blacker et al. (1989). The basic insight of the method is that a population with high fertility will, other things being equal, have a higher proportion of high-order births than a population with lower fertility. An indirect method was needed for the Beirut study because information on exposure time, specifically the numbers of women by age and religion, was not available. The method uses stable or quasi-stable population relationships to infer the underlying age distribution and derive an estimate of the total fertility rate (TFR). The study found a very small overall decline in TFR from 1984 to 1991 (from 2.6 to 2.5), with rather similar proportionate declines for Christians and Muslims, although the former have fertility 25 percent lower than the latter. The authors conclude that the war and economic hardship had little effect on fertility, despite substantial emigration of single men and the effects on marriage of a housing shortage. They attribute the small effect to the low levels of fertility already reached before the war, to migration into Beirut of higher fertility rural migrants, and to emigration of lower fertility upper-level social groups. The results of this study should be treated with caution, however. As noted earlier, indirect methods make assumptions about continuity of demographic processes that may be quite inappropriate for crisis situations. In this particular case, the distribution of births by order could be substantially affected by, for example, delayed marriage (reducing the proportion of low-order births). The method also uses an estimate of the proportion of women who remain childless, based in this case on cohort measures for women in their forties in 1984 and 1992, values determined by childbearing patterns at least 20 years earlier.

Retrospective surveys, including truncated or full-birth histories, have been used to examine the effects of civil wars, although it is generally difficult in such cases to disentangle the effects of economic conditions from those of violence or displacement. Ethiopia experienced recurrent civil wars and crop failures throughout the 1970s and 1980s. Lindstrom and Berhanu (1999) used birth history data from the 1990 National Family and Fertility Survey of Ethiopia to examine the effects of war, resettlement, famine, and economic decline on marital fertility over the period 1972 to 1988. Annual numbers of military attacks were used as indicators of conflict, and a food price index was used to identify periods of famine. The study focuses on marital fertility and thus does not examine effects of crisis on Process 1(a) effects, and it is limited to live births, so provides no indication of Process 3 effects on fertility. Checks on data quality showed no evidence of heaping on ages 0 and 5, and infant mortality rates calculated from the birth histories are close to other estimates for Ethiopia. Births were backdated to month of conception, and conception probabilities by birth order were then related to socioeconomic controls, calendar year, and indicators of crisis in the form of numbers of military attacks and dummy variables for the second year of each of four identified famines. Conception probabilities were particularly low in 1978 and low in 1980 and 1985; military attacks were high in 1978 (and 1985 witnessed a massive offensive against rebel forces in Tigray), food prices rose steeply in 1985, and both 1978 and 1985 saw steep declines in per capita gross domestic product (GDP). In a discrete time logistic regression analysis of all birth intervals, conceptions were found to be significantly reduced in the second years of famines and by the number of military attacks.

Over and above temporary declines and subsequent rebounds in marital fertility related to war, famine, or economic decline, the authors identify a steady downward trend in marital fertility after 1982, amounting to a decline of between 1.0 and 1.4 in the total marital fertility for women ages 15 to 34, attributing this decline to "the continuation of the civil wars and the deterioration of economic conditions" (p. 258). It seems likely that this apparent decline was actually the result of the use of the birth history and the common data error of pushing recent events back into the past, producing the appearance of a recent fertility decline. The 2000 Ethiopia DHS (Central Statistical Authority and ORC Macro, 2001) shows that the total fertility rate for ages 15 to 34 changed barely at all during the 1980s, averaging 5.3 for the period 1980-1984 and 5.2 for the period 1985-1989.

A second study of possible fertility effects of pervasive conflict examines the case of Angola (Agadjanian and Prata, 2002). In terms of military activity, a fragile peace lasted from May 1991 to September 1992, followed by a period of intense conflict lasting through 1993, with a new peace agreement signed in November 1994. The study used data from the 1996 Multiple Indicator Cluster Survey (MICS) of Angola. The data in this case are from a truncated birth history, collecting data on the dates of the three most recent births for each woman as well as information on fertility preferences. Also using discrete time logistic regression, birth probabilities were modeled on socioeconomic and biodemographic factors, areas of residence (categorized as the national capital area of Luanda, areas more affected by conflict, and areas less affected by conflict) and four 12-month periods from August 1992 to July 1996. In terms of main effects, more affected areas were not found to have significantly different fertility than less affected areas, but fertility was significantly lower in the two years August 1993 to July 1995 than in the years on either side. The timing of the fertility decline would correspond to a reduction in conceptions during the worst of the conflict, in late 1992 to late 1994.

The authors also estimated models including numerous interactions between area and time period, education and time period, economic condition (proxied by ownership of a radio) and time period, and parity and time period. Relative to 1993-1994, fertility increased most in 1995-1996 in Luanda and in more affected areas, among women with 5 or more years of education, and among women in households that own a radio. The data available do not permit any exploration of the intermediate variables involved in the changes. In terms of fertility preferences, at the time of the survey, women in Luanda and in more affected areas were less likely (relative to women in less affected areas) to want to get pregnant in the next 12 months, and women in Luanda were more likely to want no more children.

The authors note that the decline and subsequent rebound in fertility as fighting first intensified and then declined was most pronounced in those areas categorized as "most affected" by the war and in the capital, Luanda, that was least affected. They hypothesize that the reason why Luanda showed a strong effect is that "advanced urbanization made the residents more responsive to changes in the politico-military macroclimate and related microeconomic hardships, and, at the same time, better able to control their fertility" (p. 227). The fertility decline in the more affected areas is thus described as a largely unintentional response to social disruptions caused by the fighting, whereas the decline in Luanda is seen as the result of intentional fertility regulation. This interpretation may be stretching things somewhat: at the time of the MICS survey in 1996, modern contraceptive prevalence was only 7 percent in urban areas (and less than 1 percent in rural areas). Average numbers of children ever born by women in the 35 to 39 age group scarcely varied between urban and rural areas or between educational groups. There is not much evidence of intentional fertility regulation on a scale large enough to affect population aggregates.

In terms of data type, the authors state that they would have preferred complete birth histories but claim that recorded dates of birth and death for the three most recently born children are "generally realistic." It is not clear whether this is a statement about their own data set or about the developing world in general, but if it was meant to be the latter, it is far from correct and should not go unchallenged. Truncated birth histories, whether for the last year, the most recent birth, or for 5 years, are seriously biased (Brass et al., 1968; Bicego et al., 1991). The findings of the analysis are not necessarily distorted by possible error, but data from truncated birth histories should always be viewed with suspicion, especially when analyzed to examine changes internal to the data set over short periods of time.

It is interesting to note that the TFR estimated for Angola by the 1996 MICS was 6.9 births per woman. This figure is similar to national estimates for much of Africa, suggesting that overall fertility has not been greatly reduced by decades of war and instability.

Surprisingly little use has been made of DHS data to examine the fertility consequences of humanitarian crises. Perhaps the recognition that only surviving women get interviewed, with the attendant concern about selection bias, or the fact that refugee populations in any particular national survey are too small for separate analysis, explains this neglect of a widely available and generally high-quality source of data. These concerns are certainly valid, but large effects of population displacements might still be expected to show up.

The sudden exodus of refugees from Rwanda in July 1994, and their equally sudden return late in 1996, involved as many as 2 million people according to the high estimates; the violence before the exodus may have killed as many as 1 million people. Turmoil on this scale might be expected to show up in the birth histories in the 2000 Rwanda DHS. Figure 3(a) shows the monthly numbers of births from the beginning of 1988 to the end of 1999, and Figure 3(b) shows the proportionate deviation from a fitted trend. Note that this analysis is of numerators alone: the demographer's obsession with age and exposure time has been overcome. This is not a particularly serious problem for the analysis that follows,

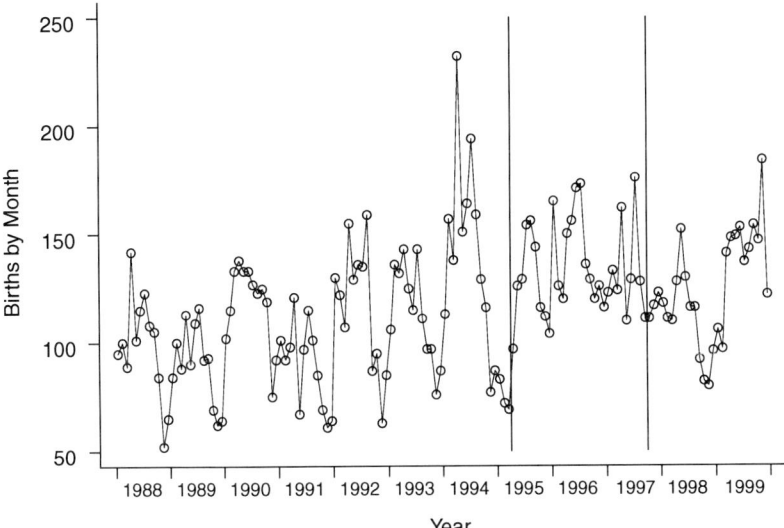

FIGURE 3(a) Births reported in Rwanda DHS by month, 1988 to 1999.

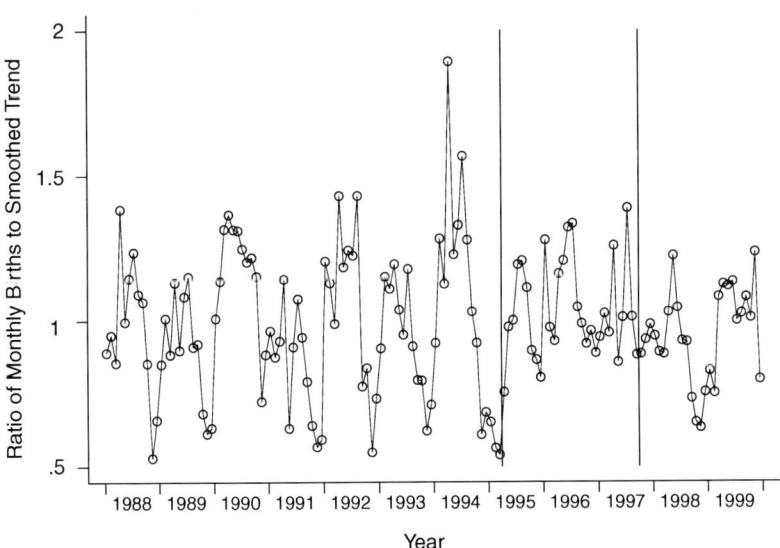

FIGURE 3(b) Proportionate deviation from trend of DHS births by month, 1988 to 1999.

since the age distribution of the reporting population is fixed at interview and cannot vary sharply over short time periods.

The data show a clear pattern for earlier years, wherein fewer births were reported in months late in the year, but there is no major decline or increase associated with the periods around March 1995 and June 1997 (marked by vertical lines on the graphs, and corresponding to conceptions in the middle of 1994 and late in 1996, when the mass moves took place), and no sudden dips that might represent periods of displacement.

The most notable feature of Figure 3(b) is the peak of births in the middle of 1994 (April to July). The following trough lasts for 5 months (October 1994 to February 1995), substantially longer than the usual end of year trough, perhaps reflecting the fertility effect of genocide prior to displacement or possibly of displacement incorrectly located in time in the birth histories. It is more likely, however, that it reflects the common error in DHS birth histories (Arnold and Blanc, 1990) of pushing birth dates about 5 years before the survey backward in time to avoid additional questions about children under age 5 (the Rwanda DHS questionnaire included additional health questions for all children born in January 1995 or later). There is no very obvious effect of the return migration, either. It is also worth noting that the survey used a monthly calendar for pregnancies, births, and other pregnancy outcomes from January 1995 onward as opposed to collecting month and year of each live birth.

It is possible that the weakening from 1995 onward of the seasonal pattern apparent prior to that year is associated with the different ways of collecting data on dates prior to and after January 1995, although there is no way of knowing which pattern (if either) is correct. Once again, the substantive conclusion is that the effect of displacement on fertility was not very great, and the methodological conclusion is that retrospective birth histories, even when complete histories, are uncertain sources of information about past fertility fluctuations, because of typical patterns of misdating of events.

Data from registration systems in refugee camps need careful evaluation. Data compiled by the Centre for Research on the Epidemiology of Disasters from camps in Ethiopia housing Sudanese refugees identified divergent fertility trends in two camps. In one, reported to have better health services, the CBR increased from about 12 per 1,000 to about 48 per 1,000 (annual estimates given here have been derived by multiplying published estimates expressed in terms of births *per month* by a factor of 12) over a 2.5-year period from early 1993, whereas in the other, reported to be more remote and to have poor services, the rate fell from about 42 per 1,000 in late 1993 to below 18 by the end of 1995 (Centre for Research on the Epidemiology of Disasters, 1997). The authors hypothesize that the rising fertility in the first camp was the result of pronatalism to repopulate the tribal group, whereas declining fertility in the second camp was the result of the harsh environment. However, doubts about the quality of the data arise from the pooled CBR for 1995 of 15.5 per 1,000 per year, and the pooled crude death rate of 3.3 per 1,000 per year. It is likely that one or more of the following factors explain the results: age distributions that are very atypical for human populations, serious underreporting of events, or serious overrecording of the base population.

Hynes et al. (2002) reviewed reproductive health indicators and outcomes based on registration systems in 52 postemergency displaced person camps in seven countries. The methodology used is of interest: one of two project investigators visited each camp and extracted information on events in the 3 months before the visit (except for maternal deaths, which were extracted for 12 previous months) from the camp registration data. The fertility indicator presented is the CBR, although the authors also present results for neonatal mortality, maternal mortality, low birthweight, and incidence of complications of abortion. They compare indicators with both the population of country of origin and the population of the host country. For the nine population groups displaced outside their country of origin, three had CBRs lower than either the origin or host country; two had CBRs lower than the host country but not significantly different from the country of origin; two had CBRs above both the host and origin countries; and two had CBRs below the country of origin, but in one case not significantly different from the host country, and in the other case above the host country. This review thus reveals little in terms of regularities of pattern in the CBR, but of course one must interpret this finding bearing in mind the weakness of this indicator as a comparative measure of fertility.

In a regression analysis of camp CBR on a range of indicators—age of camp, season, distance to border or conflict zone, time to a referral hospital, obstetric referral rate, percentage of births in the health facility, supplemental food for pregnant women, and staffing (traditional birth attendants, community health workers, and local health staff, separately)—only the age of the camp (negatively for camps in existence 7 or more years) and local health staff per 1,000 population (positively for one or more such staff person per 2,000 population) were significant. The authors note that the positive effect of local health staff may reflect better recording of births; the negative effect of camp age may also reflect a declining registration system as camps age. Across the whole sample, the CBR averaged 32 per 1,000 population per year and a percentage of the population under age 5 of 15.7; the percentage under age 5 suggests a slightly higher birth rate in a population with a normal age distribution, but one cannot draw firm conclusions given that there is no further information about the age distributions of the populations. The indicators presented look in general reasonable, although a CBR of 17 per 1,000 for Bhutanese in Nepal seems implausibly low, and a CBR of 61 for Burundis in Tanzania is unexpectedly high. Some other indicators, however, suggest omission: neonatal mortality rates of 5 per 1,000 for Somalis and 7 per 1,000 for Sudanese in Ethiopia, of 9 per 1,000 for Sudanese in Uganda, and of 7 per 1,000 for Burundis in Tanzania seem to be implausibly low.

LONG-TERM CONSEQUENCES

The Palestinian population of the West Bank and the Gaza Strip represents one of the world's largest displaced populations, and it is also the population with the longest displacement. Not only is a large refugee population involved, but also the populations of Israel, the West Bank, and the Gaza Strip have been locked in a state of belligerence for half a century. Fargues (2000) examines the effects of this state of belligerence on fertility trends in the populations involved. Data sources are problematic. Registration of births in Israel is essentially complete, but it does not identify Arabs specifically. Birth registration for Palestine may suffer from some omission, possibly varying over time as control has shifted back and forth, and estimates of the resident population are uncertain; estimates for the mid-1990s are derived from a 1995 survey.

Fargues concludes that the differences in fertility between population subgroups remained huge even into the early 1990s. Jews born in Europe and Christian Arab Israelis have had a replacement level of fertility, whereas Palestinians in the Gaza Strip have had a TFR of nearly eight. Jewish fertility has shown a pattern of convergence among different place of origin groups since 1948, whereas the fertility of Palestinians has diverged since 1970, with fertility remaining very high in Gaza, declining moderately in the West Bank, and declining rapidly (although not to Jewish levels) in Israel. Fargues argues that both sides of the conflict recognize demography as a factor in the struggle, and that both sides have adopted pronatalist policies (in Israel, specifically for Jewish fertility) to increase numbers of births and future population. Among Palestinians, many factors often associated with fertility decline (low infant and child mortality, high levels of female education, low opportunities for employment of children, a period of severe austerity) would suggest declining fertility; only low female participation in the labor force appears consistent with continued high fertility. Protracted belligerence in the region appears to have had the effect of maintaining higher fertility than would otherwise have been the case on both sides of the conflict. High fertility is also a feature of Palestinian refugees living in Lebanon (Zakharia and Tabari, 1997).

The final phase of a population displacement is return or resettlement. There is little evidence about the fertility effects of return (although the Rwanda DHS data in Figure 3 do not suggest a major effect). Some data relating to resettlement are available on the fertility of Southeast Asian populations in the United States, however. These populations had had high (Vietnam, Cambodia, lowland Lao) or very high (Hmong) fertility in their places of origin. A study by Weeks et al. (1989), based on sample surveys in San Francisco and San Diego, indicates that fertility levels among resettled refugees remained high initially, and that differentials in place of origin continued in the United States. For example, the fertility of the first largely urban-origin wave of refugees was lower than that of the second, largely rural-origin wave. The comparison of current and lifetime fertility by ethnic group for 1984 presented by Weeks et al. (1989) suggests that current TFR is actually higher than the average number of children ever born by women in their forties; this suggests the possibility of some makeup of fertility in the early years of resettlement in the United States. Evidence from studies of voluntary migrant populations generally shows quite rapid convergence of fertility toward average levels, so it is reasonable to suppose that in the somewhat longer run, fertility among refugee Southeast Asians will drop sharply.

Gordon (1989) uses data from refugee arrival records to examine the age-sex structure of the Southeast Asian refugees, drawing inferences about fertility from that structure. The age distributions (for arrivals in 1976, 1980, and 1986 of Vietnamese, Cambodians, and Laotians) are unusual in several ways: the populations were young; they included very low proportions of both elderly and very young children (with the notable exception of Cambodian arrivals in 1986, among whom 21 percent were age 5 and under); Vietnamese arrivals were predominantly young adult males; and the Cambodian age pyramids show very small cohorts of children born in 1976-1979. It is interesting to note the similarity of the patterns for Cambodians resettled in the United States and the data from the Sakaeo camp in Thailand, with the additional inference that fertility recovered very fast once the refugees reached camps in Thailand. Gordon uses child-woman ratios to examine fertility trends among arriving refugees over the period 1979 to 1986. These ratios fall sharply for arrivals from Vietnam, fluctuate around a high level for arrivals from Laos, and rise steeply for arrivals from Cambodia.

SUMMARY AND CONCLUSIONS

Data concerning the fertility impact of humanitarian crises and conflict have improved substantially over the past three decades, but they are still not adequate to answer all questions of importance.

Short-term effects can operate only through Process 3 (increased pregnancy loss) factors. There is little evidence to support a strong effect on spontaneous abortion, and evidence concerning induced abortion is largely anecdotal. Where short-term effects appear to be large (such as in the Sakaeo refugee camp in Thailand in 1979-1980) the likelihood is that the population arriving at the camp had been severely malnourished for a substantial period prior to moving. The population of Sakaeo was also selected (in terms of age and marital status) for low fertility.

Medium-term effects can also operate through Process 1 (intercourse) or Process 2 (conception) factors. The strongest data, from civil registration or surveillance data, concern the effects of famine and suggest that fertility can drop dramatically during famine, primarily as a result of reduced fecundability (although reduced coital frequency cannot be ruled out). Temporary separation must have an impact in the case of population displacement, but clear evidence for what may be a rather weak effect is lacking.

Long-term effects of conflict on fertility appear to be very limited. Populations that have experienced long-term conflicts (such as Angola) have fertility levels very similar to those of neighboring populations that have been less affected. In the Middle East, it seems likely that prolonged belligerence between Israelis and Palestinians has contributed to fertility levels higher than they would otherwise have been.

This review does not identify a uniformly satisfactory way to collect data about the demographic consequences of crisis. Sound demographic methodology, controlling insofar as possible for age and exposure effects, is important when feasible, but continuous recording of events generally breaks down in a conflict, and retrospective reporting seems to involve too much noise in the data to identify patterns.

Certain general conclusions about data collection can be drawn, however. Since age is such a crucial demographic variable, which must be controlled for to make valid comparisons, registration systems should attempt to collect and tabulate detailed information on age both for the population and for events (age at death, age of mother for births). Ideally, 5-year age groups should be used. At the very least, for fertility studies, it is necessary to record the size of the female population of reproductive age, usually taken as 15 to 44. This information will be very useful for planning the provision of reproductive health services as well. Registration systems should also ensure that enough staff are available to record events and should guard against gradual decline in quality over time. If data are to be extracted from a register for a recent period, the period chosen should be a full year to avoid possible seasonality problems. Periodic sample surveys, ideally including full birth histories, should also be implemented to provide a check on the registration system. Average numbers of children ever born by women classified by age provide a useful basis for data evaluation.

Sample surveys can provide useful fertility and reproductive health indicators from surprisingly small samples of no more than 3,000 households. A full-birth history is recommended: either time-limited or birth-limited truncation is likely to result in underestimates of recent fertility, as well as providing no basis for evaluating data quality.

A final point concerns data publication. Some reports consulted for this review provided data only in graphical form; good practice requires that the basic numbers be shown somewhere. Other papers showed summary measures but referred only to ratios of numbers for some categories. Again, interpretation may depend on having the raw numbers in such cases. Finally, I would encourage field staff conducting studies in this area to report results using standard measures, that is, to report birth and other rates in terms of person-years of exposure. Such reporting makes comparisons between studies, and between populations in crisis and those in more normal circumstances, much easier. Although it is not hard to convert a measure expressed per 10,000 population per day, or per 1,000 population per month, into the more normal format per 1,000 person-years, it is unnecessary and tends to give the erroneous impression that the measure itself is in some way inherently different, rather than just differently scaled.

ACKNOWLEDGMENT

I wish to acknowledge useful discussion and comments of participants on an earlier draft of this paper presented at the National Research Council's Workshop on Fertility and Reproductive Health in Humanitarian Crises, Washington DC, October 23-24, 2002, as well as the confidential comments of three reviewers.

REFERENCES

Agadjanian, V., and N. Prata
- 2002 War, peace and fertility in Angola. *Demography* 39(2):215-231.

Arnold, F., and A. Blanc
- 1990 *Fertility Levels and Trends.* (DHS Comparative Studies, No. 2.) Columbia, MD: Institute for Resource Development.

Ashton, B., K. Hill, A. Piazza, and R. Zeitz
- 1984 Famine in China, 1958-61. *Population and Development Review* 10(4):613-645.

Bicego, G., A. Chahnazarian, K. Hill, and M. Caymittes
- 1991 Trends, age patterns and differentials in childhood mortality in Haiti (1960-1987). *Population Studies* 45(2):235-252.

Blacker, J.G.C., M. Afzal, and A. Jalil
- 1989 The estimation of fertility from distributions of births by order. Application to the Pakistan Demographic Survey. In *IUSSP International Population Conference, New Delhi.* Liege, Belgium: International Union for the Scientific Study of Population.

Bongaarts, J.
- 1977 A dynamic model of the reproductive process. *Population Studies* 31(1):59.
- 1980 Does malnutrition affect fertility? A summary of evidence. *Science* 208:564-569.
- 1982 The fertility-inhibiting effects of the intermediate fertility variables. *Studies in Family Planning* 13(6/7):179-189.

Brass, W.
- 1969 Disciplining demographic data. In *Proceedings of the IUSSP International Population Conference, London.* Liege, Belgium: International Union for the Scientific Study of Population.

Brass, W., A.J. Coale, P. Demeny, D.F. Heisel, F. Lorimer, A. Romaniuk, and E. van de Walle
- 1968 *The Demography of Tropical Africa.* Princeton, NJ: Princeton University Press.

Brass, W., and H. Rashad
- 1992 Evaluation of the reliability of data in maternity histories. In A. Hill and W. Brass (eds.), *The Analysis of Maternity Histories.* Liege, Belgium: International Union for the Scientific Study of Population.

Centers for Disease Control
- 1983 International notes surveillance of health status of Kampuchean refugees—Khao-I-Dang Holding Center, Thailand, December 1981-June 1983. *Morbidity & Mortality Weekly Report* 32(31):412-415.

Central Statistical Authority and ORC Macro
 2001 *Ethiopia Demographic and Health Survey 2000*. Addis Ababa, Ethiopia, and Calverton, MD: Central Statistical Authority and ORC Macro.

Centre for Research on the Epidemiology of Disasters
 1997 *Reproductive Health Needs of Refugees: Evidence from Three Camps in Ethiopia*. Brussels, Belgium: Centre for Research on the Epidemiology of Disasters, Department of Public Health, Université Catholique de Louvain.

Chowdhury, A.K.M.A, and L.C. Chen
 1977 *The Dynamics of Contemporary Famine*. (Report No. 47.) Dacca, India: Ford Foundation.

Curlin, G.T., L.C. Chen, and S.B. Hussain
 1976 Demographic crisis: The impact of the Bangladesh civil war (1971) on births and deaths in a rural area of Bangladesh. *Population Studies* 30(1):87-105.

Davis, K., and J. Blake
 1956 Social structure and fertility: An analytic framework. *Economic Development and Cultural Change* 4(4):211.

Fargues, P.
 2000 Protracted national conflict and fertility change: Palestinians and Israelis in the twentieth century. *Population and Development Review* 26(3):441- 482.

Galloway, P.
 1988 Basic patterns in annual variations in fertility, nuptiality, mortality and prices in pre-industrialized Europe. *Population Studies* 42:275-303.

Gordon, L.W.
 1989 The missing children: Mortality and fertility in a Southeast Asia refugee population. *International Migration Review* 23(2):219-237.

Holck, S.E., and W. Cates, Jr.
 1982 Fertility and population dynamics in two Kampuchean refugee camps. *Studies in Family Planning* 13(4):118-124.

Hynes, M., M. Sheik, H.G. Wilson, and P. Spiegel
 2002 Reproductive health indicators and outcomes among refugee and internally displaced persons in postemergency phase camps. *Journal of the American Medical Association* 288(5):595-603.

International Centre for Migration and Health
 no Pregnancy outcome among displaced and non-displaced women in Bosnia and
 date Herzegovina. In *Report of the Technical Working Group on Reproductive Health and Pregnancy Outcome Among Displaced Women*. Geneva, Switzerland: International Centre for Migration and Health.

Khlat, M., M. Deeb, and Y. Courbage
 1997 Fertility levels and differentials in Beirut during wartime: An indirect estimation based on maternity registers. *Population Studies* 51:85-92.

Lee, R.
 1990 The demographic response to economic crisis in historical and contemporary populations. *Population Bulletin of the United Nations* 29:1-15.

Lindstrom, D.P., and B. Berhanu
 1999 The impact of war, famine and economic decline on marital fertility in Ethiopia. *Demography* 36(2):247-261.

McGinn, T.
    2000 Reproductive health of war-affected populations: What do we know? *Family Planning Perspectives* 26(4):174-180.

Médecins Sans Frontières
    1997 *Refugee Health: An Approach to Emergency Situations.* New York: MacMillan Education LTD.

Mosley, W.H.
    1978 The effects of nutrition on natural fertility. In H. Leridon and J. Menken (eds.), *Natural Fertility.* Liege, Belgium: International Union for the Scientific Study of Population.

Palloni, A., K. Hill, and G.P. Aguirre
    1996 Economic swings and demographic changes in the history of Latin America. *Population Studies* 50:105-132.

Palmer, C.A., L. Lush, and A.B. Zwi
    1999 The emerging international policy agenda for reproductive health services in conflict settings. *Social Science and Medicine* 49:1689-1703.

Preston, S.H., P. Heuveline, and M. Guillom
    2001 *Demography: Measuring and Modeling Population Processes.* Oxford, England: Blackwell Publishers.

Stanton, C.
    in press Methodological issues in the measurement of birth preparedness in support of safe motherhood. *Evaluation Review.*

Stein, Z., M. Susser, G. Saenger, and F. Marolla
    1975 *Famine and Human Development: The Dutch Hunger Winter of 1944- 1945.* New York: Oxford University Press.

United Nations
    1983 *Manual X: Indirect Techniques for Demographic Estimation.* (Population Studies, No. 81, Department of International Economic and Social Affairs.) New York: United Nations.
    1994 *Report of the International Conference on Population and Development, Cairo, 5-13 September 1994.* A/CONF.171/13/Rev.1. Available: http://www.unfpa.org2/icpd/icpd_poa.htm [March 31, 2004]
    1996 *Population and Development.* New York: United Nations.

Weeks, J.R., R.G. Rumbaut, C. Brindis, C.C. Korenbrot, and D. Minkler
    1989 High fertility among Indochine refugees. *Public Health Reports* 104(2):143-150.

Wulf, D., ed.
    1994 *Refugee Women and Reproductive Health Care: Reassessing Priorities.* New York: Women's Commission for Refugee Women and Children, International Rescue Committee.

Zakharia, L.F., and S. Tabari
    1997 Health, work opportunities and attitudes: A review of Palestinian women's situation in Lebanon. *Journal of Refugee Studies* 10(3):411-429.

ANNEX:
RAW NUMBERS UNDERLYING FIGURES 1 TO 3

Raw Numbers Underlying Figures 1 to 3

Number of Births

Month of Birth	1988	1989	1990	1991	1992	1993	1994	1995	1996	1997	1998	1999
January	95	84	102	101	130	106	113	83	165	123	118	106
February	100	100	115	92	122	136	157	72	126	133	111	97
March	89	88	133	98	107	132	138	69	120	124	110	141
April	142	113	138	121	155	143	232	97	150	162	128	148
May	101	90	133	67	129	125	151	126	156	110	152	149
June	115	109	133	97	136	115	164	129	171	129	130	153
July	123	116	127	115	135	143	194	154	173	176	116	137
August	108	92	123	101	159	111	159	156	136	128	116	143
September	105	93	125	85	87	97	129	144	129	111	92	154
October	84	69	119	69	95	97	116	116	120	111	82	147
November	52	62	75	61	63	76	77	112	126	117	80	184
December	65	64	92	64	85	87	87	104	116	123	96	122

SOURCE: Rwanda DHS 2000: Births by Month and Year.

Pregnancy Outcomes by Month: Matlab Demographic Surveillance System

	Miscarr.	Still-births	Live Births	Miscarr.	Still-births	Live Births	Miscarr.	Still-births	Live Births	Miscarr.	Still-births	Live Births
January	34	33	1,067	32	28	838	63	26	614	78	45	1,188
February	44	34	940	31	21	711	71	22	577	76	37	1,009
March	31	20	931	50	15	695	93	27	638	91	38	973
April	38	18	811	42	22	603	93	29	501	94	36	813
May	42	35	791	50	16	466	108	29	668	123	31	739
June	54	34	743	57	23	347	103	29	752	94	48	801
July	46	29	811	52	18	490	117	36	898	105	40	886
August	48	22	805	50	16	586	114	39	1,013	84	36	1,042
September	50	34	1,004	65	18	795	79	64	1,281	71	49	1,146
October	31	44	1,128	57	18	780	73	57	1,367	75	63	1,328
November	31	33	1,155	61	30	637	68	55	1,487	75	59	1,352
December	29	41	1,027	71	26	613	57	56	1,375	48	47	1,111

ABOUT THE AUTHOR

Kenneth Hill has been a professor and director of the Hopkins Population Center since 1995. His research interests have been in the development of demographic measurement methods (particularly for demographic outcomes that are hard to measure, such as child and adult mortality, unmet need for family planning, undocumented migration); the measurement of child mortality (with particular emphasis on tracking national trends and linking them to other changes); the exploration of links between demographic parameters and economic crisis; the impact of policy and programs on demographic change; the role of gender preferences on child health behaviors and fertility; the demography of Sub-Saharan Africa; the role of development, particularly child mortality change, on fertility decline; and the measurement of demographic parameters for populations undergoing complex emergencies. His publications include "Demographic Techniques: Indirect Estimation" (*International Encyclopedia of Social and Behavioral Sciences*, 2001); "Interrupting HIV Transmission in Africa" (*Foreign Policy*, 2001); *Levels and Trends in Child Mortality in the Developing World* (with R. Pande, 2001); *The Quick and the Dead in Zimbabwe: Replacement and Insurance Effects on Fertility* (with R. Marindo and M. Mahy, 1997); "Demographic Responses to Economic Shocks: The Case of Latin America" (with A. Palloni, in *The Peopling of the Americas*, Vol. 3, 1992). He has a Ph.D. in demography from the London School of Hygiene and Tropical Medicine.

The **Committee on Population** was established by the National Academy of Sciences (NAS) in 1983 to bring the knowledge and methods of the population sciences to bear on major issues of science and public policy. The committee's work includes both basic studies of fertility, health and mortality, and migration and applied studies aimed at improving programs for the public health and welfare in the United States and in developing countries. The committee also fosters communication among researchers in different disciplines and countries and policy makers in government and international agencies.

The **Roundtable on the Demography of Forced Migration** was established by the Committee on Population of the National Academy of Sciences in 1999. The Roundtable's purpose is to serve as an interdisciplinary, nonpartisan focal point for taking stock of what is known about demographic patterns in refugee situations, applying this knowledge base to assist both policy makers and relief workers, and stimulating new directions for innovation and scientific inquiry in this growing field of study. The Roundtable meets yearly and has also organized a series of workshops (held concurrently with Roundtable meetings) on some of the specific aspects of the demography of refugee and refugee-like situations, including mortality patterns, demographic assessment techniques, and research ethics in complex humanitarian emergencies. The Roundtable is composed of experts from academia, government, philanthrophy, and international organizations.

Other Publications of the Roundtable on the Demography of Forced Migration

Psychosocial Concepts in Humanitarian Work with Children: A Review of the Concepts and Related Literature (2003)
Initial Steps in Rebuilding the Health Sector in East Timor (2003)
Malaria Control During Mass Population Movements and Natural Disasters (2003)
Research Ethics in Complex Humanitarian Emergencies: Summary of a Workshop (2002)
Demographic Assessment Techniques in Complex Humanitarian Emergencies: Summary of a Workshop (2002)
Forced Migration and Mortality (2001)